Numerology for Beginners

The Simple Guide to Discovering Numbers That Resonate With Your Relationships, Future, Money, Career, and Destiny

Donald B. Grey

Bluesource And Friends

This book is brought to you by Bluesource And Friends, a happy book publishing company.

Our motto is **"Happiness Within Pages"**

We promise to deliver amazing value to readers with our books.

We also appreciate honest book reviews from our readers.

Connect with us on our Facebook page www.facebook.com/bluesourceandfriends and stay tuned to our latest book promotions and free giveaways.

Table of Contents

Introduction

From the moment of your birth, to that time when someone asked you how old you were and you held up three little fingers to show your age, numbers have played an important role in your life.

Do you remember waiting impatiently to turn thirteen? This event was the gateway to your teenage years, which would be filled with a great deal of freedom, friendship, and carefree fun. Then sixteen was your next exciting birthday, when you could apply for your driver's license. And, shortly thereafter, you celebrated your twenty-first birthday, where you were considered an adult who could legally order a drink at a bar. Now, that was a day to remember! You were suddenly emancipated from your childhood dependence, and capable of running your own life!

So, as you can already see, numbers, even if these relate only to birthdays at this point, have had a great deal of influence in your life.

What Is Numerology?

You may not have thought about the importance of numbers beyond what you required for math class. You may then be amazed to know that number combinations can prove to be useful tools for divining your future, relationships, potential business success, health, wealth, and your destiny. If this idea sounds a little far-fetched, please read on to discover more about this ancient channel of prediction.

Numerology, understood as the universal language of numbers and the first cousin to astrology, is a global way of communicating using numbers to confirm facts about personal details and the path you have chosen in life. When correctly interpreted, numerology can give you useful insights into your personality, the most suitable career, your potential wealth and well-being as well as your health and longevity.

Numerology can be described as the study of the relationship between people and numbers, as well as how these numbers all relate to each other (Dworjan n.d.).

The concept of numerology suggests that the cosmos, as well as every aspect of your life, is made up of and controlled by basic elements consisting of numbers. In this case, your life has been mapped out according to your date of birth, the name you were given at your birth, and a number of other important numerical factors (Hurst, 2020).

Through our knowledge that everything in the universe can be broken down into elements of which numbers make up the basic unit, a numerologist is able to assign meaning to your life——including your past, present, and future.

There are 11 important numbers in numerology, ranging from 0 to 11. Each of these numbers relates to a specific stage or cycle in your life and provides insight into your nature, personality, and destiny.

Numerology can be used to predict certain future incidents, as well as interpret important aspects about your character and how you can improve areas that are less than perfect. However, the rewards of numerology are directly proportional to the amount of effort you are prepared to invest into learning about it.

What is a Numerologist?

Numerologists understand and equate everything in our universe to numbers. They are then able to isolate and identify individual numbers that have value to you as well as those that may be harmful. Through careful study and in-depth calculations of your personal numbers, numerologists can help you to define your personality and discover your true purpose in life.

The History of Numerology

"Numerology" is a term coined by Dr. Julian Stenton, and, though this study of numbers can be closely linked to mathematics and is most likely to have originated during the times of the ancient Greeks and Romans, there is also evidence of its existence in early written records from Babylonia and Egypt (Hurst, 2020).

Amongst the Babylonians was a group called the Chaldeans, who were astronomers and astrologists who used complicated calculations by using a person's given name to divine their future. The Chaldeans used the digits 1-8, divided the alphabet into sets of seven letters, each of which corresponded with the sun, moon, Mercury, Mars, Venus, Jupiter, and Saturn.

Using the alphabetical letters in conjunction with the planets, the Chaldeans made predictions about people's futures, wealth, and destiny (Kali, 2019).

The Ancient Hebrews also used arithmancy or numerology to decode hidden messages in the Bible to discover its sacred power and energy.

Pythagoras, an ancient Greek philosopher and mathematician, noticed that numbers work in cycles. He believed that the foundation of reality in the universe was mathematical in nature. His great interest in numbers led him to develop a theory for calculating the sizes of the sides in a right-angled triangle, $C^2 = A^2 + B^2$.

The Fibonacci Theory, another interesting number hypothesis, is closely related to Binet's golden ratio in which the *nth* number in the Fibonacci sequence is expressed as n. The golden ratio is denoted by the sum of two preceding, consecutive, or sequential numbers that add up to a third number (Fibonacci Number, 2020).

This mathematical sequence creates a spiral-shaped helix, similar to that found in the shell of a mollusk. It also appears in computer algorithms as well as in nature, where it can be determined from the pattern of branches in a tree or leaves on a twig. The same sequence is also evident in the fruit sprouts on a pineapple, the bracts on a pine cone, and the leaves of an unfurling fern.

Unfurling Fern Leaves. From Unsplash, Uploaded by Lenor Barry (n.d.), https://unsplash.com/s/photos/fibonacci-in-nature

The sequence, denoted by F, reads as follows: F1 added to F1 equals F2. Subsequently, F1 added to F2 equals F3, and F2 added to F3 equals F5.

Supporting the mathematical importance of numerology is the fact that every name is assigned a value with a specific numerical vibration that can be measured. Essentially, each number holds secret clues to the true nature of people, places, animals, and objects (Kali, 2019).

The Link Between Numerology and Astrology

As in astrology, where zodiac signs are linked to the planets and suggest specific characteristics and traits in people, so too does numerology use numbers and number combinations to discover information about human personal aspects.

Although both astrology and numerology are viewed by some as pseudoscience, they can work hand-in-glove to create an awesome combination to determine our destiny.

How Numerology Actually Works

Working with numbers may sound like a straightforward activity, but when the true worth of numerology is considered, it's the interpretation of the interconnection between specific numbers that give your special numbers their authentic value. Analogous to your horoscope, numerology can be used to interpret signs and numerals to determine your future. It is therefore possible to assume that there are no coincidences in life, and that the path you choose and the personality you develop are predetermined by your date of birth and full birth name.

A numerology reading takes time and involves a great many calculations using different combinations of numbers to arrive at specific conclusions. Because numbers are infinite, your numerology reading can continue until the end of your lifetime, revealing dozens of different aspects of your life, personality, and the changes that come with age and progression.

In order to discover your true destiny, let's take a look at the power of your personal numbers in your life and how these influence everything you do, say, and think.

Chapter 1: The Power of Your Personal Numbers

Numerology discloses the age and essence of your soul. It is a useful tool for making sense of recurring numbers that may impact your life in a number of ways. The repetition and synchronicity of numbers have been observed and studied for thousands of years with the view of discovering their true meaning and how they can influence our present and future (Faragher, 2020).

When you discover the personal numbers that resonate with you, creating a frisson in your core, you will be able to effectively manipulate the aspects of your life to ensure you make the best decisions at work, understand understand compatibility with your partner, choose successful pursuits that will lead to increased happiness, and find opportunities to realize your true destiny.

Your personal numbers are those that are formed from the year, month, and day of your birth. These numbers may have positive or negative influences on your life, happiness, and future success. There is, however, no denying the important role they play in shaping your destiny.

With knowledge of your personal numbers, you may have the power to challenge certain incidents and control how these play out, instead of being caught by surprise. Clearly, knowledge of your numbers can arm you against potential dangers and pre-warn you of impending disasters.

Many well-known celebrities admit to using the power of numbers in their daily lives to foretell the future, help them make decisions, and even control current events in their lives. Among these celebs is Wynona Ryder, who is a staunch believer in the power of numbers (The Public Figures Who Believe in the Mystical Significance of Numbers, 2011).

Kim Kardashian's life path number is 4, which means she is well-grounded with a strong awareness of how best to handle situations to her benefit. Her orderly, organized, and decisive nature makes Kim a practical, down-to-earth person with strong ideas about right and wrong. Kim has a rational approach to solving problems and seldom gives up readily once she commits to something she believes in. She employs hard work and consistent effort to achieve her goals without relying on unconventional short cuts. Kim believes that honesty, integrity, and justice form the pillars of society (Tommer, n.d.).

The Power of Numbers

The study of numerology is a self-help tool that can be utilized to your benefit. The power of numerology can help you gain better insight into yourself, and empower you to overcome many obstacles. With knowledge of your personal numbers and how these reflect your past, influence your present circumstances, and have the power to influence your future, you have the tools to manipulate your life by uncovering truths that would otherwise remain hidden.

What Your Chart Indicates

Your personal numerology chart will give you valuable information on your past and how to rectify responses to past mistakes, so that you may be more at be more at peace. It will also assist in defining your future and suggest changes in your behavior to support improvements in your health, relationships, wealth, inner happiness and peace.

Your number chart gives advice and support for present circumstances and how these can be positively manipulated in order for you to smooth the way forward to a better future. The chart shows your strengths and weaknesses, talents, gifts, and potential challenges that you will have to face. Your number chart can guide you towards making more favorable relationships and strengthen your resolve against negative circumstances. Every number in your chart carries a great deal of information about your true potential, weaknesses, and ways to overcome these. It is up to you to decide how much or how little your numerology chart will affect your life.

In order to really benefit from numerology, you should commit to allowing yourself to open up to the insights your numbers suggest and make a concerted effort to change things in order to live your best life. You only have one life, and every moment you're given is a moment lost if it's not used.

Remember, as with all things, balance is of optimal importance to achieving your true success. Your preferences to specific numbers will have more influence in your life than other numbers. These numbers will then become a part of your intricate number chart and may then play a significant role in your life.

Cairn Stones at Skogafoss Falls, Iceland. From Unsplash, Uploaded by Martin Sanchez (n.d.), https://unsplash.com/s/photos/balance

The Core Elements of Numerology

The five core elements considered to be the foundation of the direction in your life include your Birthday and Soul Urge Numbers, Personality and Expression Numbers, and your Life Path Number.

Your Expression Number

This important number is linked to your destiny and gives an indication of your desires, goals, and inherited traits.

Use the Pythagorean Number Chart to discover your Expression Number by converting each letter of your full birth name to digits. It is essential that you use the names given to you at birth in order to find your true Expression Number.

Pythagorean Chart

1	2	3	4	5	6	7	8	9
A	B	C	D	E	F	G	H	I
J	K	L	M	N	O	P	Q	R
S	T	U	V	W	X	Y	Z	

Once you have the total of your full name, reduce the number to its smallest digit. This final number works in conjunction with your Life Path Number to indicate who you really are.

Your Soul Urge Number

This number is a reflection of your inner self, the self that people seldom see. It reveals the truth about which only you know. For example, you may have a deep desire to wield power over others, or to be accepted and affirmed.

Knowledge of your Soul Urge Number can give you a new and exciting direction in life. It can also help you figure out how best to deal with your inner demons and lay them to rest.

To calculate your Soul Urge Number, you should add the value of the vowels in the name you were given at birth. Then reduce the answer to a single number. The final number may turn out to be 11 or 22, which are Master Numbers.

Your Personality Number

This useful number shows the side of you that other people see. It relates to your public persona and it can be quite enlightening to discover those traits in your nature that you take for granted and which may need some serious attention for improvement.

Your Personality Number is calculated by adding all the consonants from the name you were given at birth. Once you arrive at a total, reduce this to a single number. If the final number is an 11 or 22 it cannot be reduced, as it is a Master Number.

The Number of the Day of Your Birth

This number is the exact day on which you were born. When combined with all your other numerology numbers, your Birthday number gives a clear indication of your destiny and your life's purpose, in conjunction with your talents and gifts.

Taken together, your numbers have the power to give you direction in life, heal old hurts, govern your future successes, and lead you to improved health, vitality, and inner peace. These numbers can also be indicative of your financial situation and give you guidance on how to increase your wealth.

The Milky Way. From Unsplash. Uploaded by Maarten
Verstraete (n.d.), https://unsplash.com/s/photos/astronomy

Your Personal Year Number

Each year brings its own changes and energies for your
potential renewal, growth, and development.

Your personal year numerology chart spans nine years.
Once this nine-year cycle ends, a new cycle begins. The
very nature of change makes it possible for everything
in the universe to remain in constant motion. Your life
is not static, but in a constant state of flux. Because of
the continued change in your life, numerology is used
to predict the coming year while also giving you
valuable information about what has already happened
during the past months.

The value of learning your year number is so you can be as well prepared as possible for what lies ahead. Through careful and consistent planning, you can avoid potential pitfalls and maximize the positives to your advantage. By being able to forecast your future, you begin to exercise power over how it will turn out.

Your Personal Month Number

The energy embodied by your Personal Month number will affect the events and occurrences you may experience during the month. This number is derived from the number of your personal year, added to the number of your personal month. The total is then reduced to a single digit. So, if your personal year number is 8 and the month of your birthday is September, $8 + 9 = 17$. Further, reduce 17 to $1 + 7 = 8$, and will discover and you will derive your personal month number.

If you have information about what the month holds for you, your chances of planning accordingly will improve.

Your Personal Day Forecast

Similarly, your Personal Day forecast gives you added information about the best plans for the day. If we take the number of the day to be 3, the following applies:

This is the perfect day to entertain, have fun, and relax with good friends. Allow your creative side to show its true worth through painting gorgeous pictures, practicing on your favorite instruments, or starting on the book you want to write. A number 3 day is also a great time to gift resources to or spend time with to or spend time with those less fortunate than yourself (Bender, 2020).

Crystal Ball. From Unsplash. Uploaded by Michael Dziedzic (n.d.), https://unsplash.com/s/photos/fortune-telling

Your Personal Year Forecast

This number is deduced by adding your date and month of birth to the current year. So, as per above, 12 December will remain as 3 + 3, while the year will be 2020, which reduces to 4. Added together, your personal year number is, 3 + 3 + 4 = 10. This number is now reduced to 1.

Your personal year number sets the stage for the entire year. It is in fact like your zodiac sign, but this time in number format, and it endures for the entire year. So, although your zodiac sign is Sagittarius, and your Chinese year sign is the year of the snake, your numerology number is 1.

A personal year with a 1 signifies the start of something new. You are on the threshold of a new beginning. You have the chance to begin something great and to succeed. Trust in yourself and stand firm in your belief that you will succeed.

The all-encompassing theme for your personal year spells independence. This period indicates the start of a new nine-year cycle. You should now cultivate the seeds of what you want to achieve in the long-term.

This is a busy and tiring year, but you should stay focused on your resolve to plan and lay the foundations for the change you want to see happen in your life. You may find you feel lonely during this year. However, remain steadfast to your end goals. Ensure that you will stop anyone from taking advantage of you and your resources. Encourage others to be independent of you. The year 2020 is a universal year 4, which signifies hard work, and staying grounded will benefit you in the long run.

You may go through a profound energy shift during the year, as the previous cycle closes and the new one begins. Use this year to begin a new project and to build your future. It is all about planning and staying focused, keeping your eye on your end goal so that you may do all in your power to reach it.

Your life path number and personal year forecast run hand-in-hand to ensure an all-encompassing prediction for your ultimate good (Bender, 2020).

The Value of Your Special Numbers

Number 1

This is the number that suggests a new beginning. Perhaps, number 1 may allude to a new relationship or friendship. It could denote a fresh start in a business venture or a move to a different town.

Number 1 brings abundant blessings and relief from past traumas, illness, and pain. It foretells of good health, increased opportunities, and the accompanying peace of mind these positive changes bring.

When there is a sudden strong display of 1s in your life, the universe is encouraging you to make a fresh start. You may become motivated to try something new or embark on an adventure you may never have considered otherwise.

Change Neon Light Signage. From Unsplash, Uploaded by Ross Findon (n.d.),https://unsplash.com/s/photos/zodiac-signs

Once you accept the presence and meaning of number 1, you will begin to show your leadership skills by confidently taking control of situations and accomplishing tasks diligently and timeously.

Those around you are likely to notice the change in your demeanor and accord you praise for being brave and forthright in your approach to your life. Now is the time to extend yourself and confidently request consideration for a raise or promotion.

A young friend who had suffered horrific abuse at home suddenly discovered an interest in numerology. From the get-go, she realized that the number one kept popping up in a variety of ways. Finally, she consulted with a well-known numerologist who suggested that perhaps the universe was giving her a nudge to move on and start afresh. Some years later, we met up and she explained how fortunate she had been to follow that advice. She now has a successful real estate business, and the number of her beautiful new home is 820, which reduces to 1!

Number 2

Twos represent partnerships and joint ventures. Marriage may be in the cards if you are in a relationship. However, do not discard the possibility of forming a working relationship that may prove lucrative as well as intellectually stimulating.

When 2s make themselves known in your life, there is a greater chance of cooperative interaction. You may also have the opportunity to choose between two jobs, two people, or two potentially viable opportunities.

Now is the time to improve your communication skills and allow yourself the opportunity to step forward and have your say. The number 2 offers you increased stability, balance, and a significant decrease in your stress level.

A family member who had fallen on hard times found 2s appearing in his life. Not being aware of their significance, he mentioned this phenomenon when he visited me one day. Having explained the potential value of consistently noticing 2s in his environment, I asked if he had any plans for his future. He shyly admitted he was interested in purchasing a local gym. Although he had plenty of business knowledge, a lack of finances had held him back from taking the next step. After some discussion, I suggested we go into business as partners, and neither of us had looked back since we opened our gym 15 years ago.

Number 3

If three is presently a dominant number in your life, your innate creativity is ready to be put into action. Previously challenging situations may now melt away under your clever, optimistic, and enthusiastic manipulation of events.

Three inspires you to repaint the house, redecorate that drab sitting room, or revitalize your garden. You may decide to take up art, pottery, or even ballroom dancing when 3s dominate your life.

The only drawback with three is the development of the eternal triangle where two of the parties may challenge the third. In this event, keep an open mind and use your creative problem-solving skills to find a workable solution.

My mom and I faced the challenge of renovating our much-loved, but very tired-looking, home. Neither of us knew where to begin and so project after project was put on the back burner. Some weeks ago, I discovered the number three being more evident in my life. Three long-standing friends, Roy, Jan, and Tom, invited me out to celebrate our 12 years out of college. They had booked a table at a restaurant with a street number 7302 and we sat at table 3.

Realizing the potential value of the repetitive three, I immediately thought about all the home projects that needed to be done. Within a few weeks, the house was looking great. Even my sister, who potentially saw the negative side of every challenge, was on board for the project. Mom gave my sister the responsibility of sourcing for all the supplies, and our threesome worked out pretty well.

Number 4

Soon after the house was redecorated, we discovered the evidence of plenty of 4s. With relief, I realized that the constant reminder of number 4 was a sign of stability. Four is a fully grounded number, resting solidly on its foundation. Just as one of the sturdy old wooden chairs we had refurbished that stood proudly in our redecorated living room, Number 4 offered solidarity and security.

Now that we had taken care of all the nagging details and restored our home to its true glory, we were able to enjoy the fruits of our labors resting in the sure knowledge that now that everything was mended, our home will offer us security, stability, and peace of mind.

Number 5

Number 5 signifies change and the incumbent stress associated with any upheaval and break from routine. You may fear change, as many people do, but once you have decided to take the next step into the future, you usually feel better (Bakula, 2016).

Change can be beneficial if you find yourself tired of the same old routine, and yearning for something new.

A smaller property came onto the market recently. The house number was 707. Out of interest, mom and I took a look at the house and immediately liked its situation-,—just on the outskirts of town, close to a little river. After a great deal of discussion and planning, we sold our large old family home-,—yes, the one we had renovated, and moved into the cute cottage at number 707. What a windfall! 5s can bring good opportunities for change.

Number 6

Abundant 6s are the sign of hard work and industrious effort. It's time for immediate action. Usually, 6s like to be around groups of people, but you may find yourself becoming more active in a specific community project because 6 is at work in your life.

Some months ago, our small community group of six families, which we jokingly named """"The Busy Bees"""", decided on a specific project to help homeless people in our small town. On the 6th day of June 2004, we set out on our adventure. By the end of that same year, on the 12th day of December, we had found homes for 6 families and jobs for 6 homeless adults.

Exciting things happen when you take note of the messages the universe sends you!

Number 7

The number 7 stands for victory, but only hard-won triumph after a challenge of some sort. Seven can also signify that someone else wants what you have or you may find yourself feeling dissatisfaction with your job, relationship, or situation in life. Either way, change is in the offing. The challenge, however, is to ensure you remain patient and focused until the right time comes to make the change.

Seven also signifies the need to search for truth and a deeper spiritual understanding. Because 7 is a sacred number found in scripture (the world was created in 7 days), in nature (the rainbow has 7 colors), it is a trustworthy, reliable number. Your instincts at this time when 7s make themselves known to you will lead you to success and a higher understanding of the universe and its treasures (Bakula, 2016).

Number 8

The number 8 signifies power over your finances, wealth, and possessions. Often 8s are alone in their endeavors, so you will need to remain strong and focused if 8 begins to show up in your life.

8 suggests balance both above and below, and carries the message that nothing worthwhile comes without hard work and focused effort. People who have 8s in their lives are often successful and satisfied with their lot in life.

Number 9

The number 9 brings you closer to the end of your journey, whether this is of a personal or business nature. An influx of 9s in your life warns you of the need to cut ties with anything or anyone that holds you back or acts as a deterrent to your future happiness (Bakula, 2016).

It is a time to discard old ideas and experiences and look forward to an opportunity for aa spiritual and mental rebirth and renewal by recharging your body, mind, and soul.

Number 9 also signifies that the end of a project is in sight. Perhaps you have been busy with a specific activity that has cost you many hours of effort. When 9s begin to show up in your numbers, the time for task completion is near.

A Final Word

Your life and happiness are governed by a series of special numbers, each of which relates to you in a very specific way.

Once you realize the value of these numbers, you gain the power to manipulate this knowledge to add value to your life and guide yourself towards a well-deserved success and happiness.

Now that you have a good inkling of the value of your specific numbers, are you ready to check out how these numbers impact your personal life, love, and passion? Then read on!

Chapter 2: The Impact of Numbers on Your Personal Life, Passion, and Love

Numerology is an interesting phenomenon that becoming more popular with people from all walks of life. Being able to recognize and decode messages received through the universal medium of numbers, as well as understanding the impact these numbers have on our lives, can certainly give us an advantage when it comes to planning our personal future as well as setting our business goals (Holly, 2016).

By paying attention to numbers that come into your personal orbit, you allow yourself to open up to the opportunities that are being presented.

Zodiac-Signs. From Unsplash, Uploaded by Josh Rangel
(n.d.),https://unsplash.com/s/photos/astrology

Understanding Important Numbers

As mentioned earlier, numbers carry specific values. Besides each of these numbers being worth a fixed amount in numerical terms, each number carries a vast capacity of meaning for your life, relationships, and passion.

Because numbers vary in value, the energy emitted by each number will differ. The important thing is for you to discover your personal numbers, and then put the predictions into action.

Through numerology, you can use this information to take control of your loves, relationships, and passions in order to recreate happiness and fulfillment in your life. So, you will use your Life Path Number to discern where your relationships will head this year. This revelation will set you on the best course for success.

The Number That Designates Your Life Path

Your Life Path Number plays a vital role in every aspect of your numerology chart. It equates to your astrological sun sign. Its value lies in its power to determine your characteristics, strengths, weaknesses, aspirations, and talents. Your Life Path number shows your past, present, as well as your future.

To calculate your Life Path number, reduce the date, month, and year of your birth to a single digit and add ,them, and again reduce the answer to a single digit (Faragher, 2020).

For example, it you were born on the 12 December 1989, the calculation will look like this:

Date: 12 = 1 + 2 = 3

Month: December is the 12th month so 1 + 2 = 3

Year: 1989 = 1 + 9 + 8 + 9 = 27; 2 + 7 = 9

By adding 3 + 3 + 9 you will discover that your Life Path = 6.

Life Path Number 1

If you have a Life Path 1, you are a natural leader -,— the 'alpha' member in your home and work. You generally take control of things, and because your independent nature leads you to be an achiever, you are not afraid to work hard in order to achieve your goals. However, because of your innovative nature, details soon bore you. You prefer to create and manage tasks, and are happy to employ others to do the menial work.

You mix well in social groups and are generally well-liked. You can be assertive, but you struggle to take instructions. Your ambition drives you relentlessly forward to achieve your goals (Weist, 2019).

Famous People

Other well-known people with Life Path 1 include Martin Luther King Jr, Tom Hanks, and Sting (Laine, n.d.).

Your Relationships

If you want to establish a closer and more intimate relationship with another you need to learn to listen more attentively to your partner and express your love more frequently. Finding a good balance between work and your love life will be beneficial.

Life Path Number 2

If you have a Life Path 2, you are a balanced person who is destined to be a peacemaker and healer. Your diplomatic qualities make you an excellent mediator. You are highly intuitive and idealistic. Your perfectionist nature means that you enjoy routines and are good at planning in all areas of your life.

You find it difficult to work and remain peaceful in discordant situations, preferring to step in between warring parties to re-establish peace. Your extreme empathy and care for others could lead to your downfall. Guard against losing your emotional energy in non-productive situations (Weist, 2019).

Famous People

Jennifer Aniston, Meg Ryan, and Jackie Kennedy are famous people with a Life Path 2. (Laine, n.d.).

Your Relationships

If you are in a relationship, nurture this bond with care and empathy. This is the year to explore new opportunities and share some truly memorable occasions with your partner. Because you prefer a partner you can talk to and interact with on a mental, spiritual, and physical level in order to develop strong bonds together, you may still be waiting for the right person to cross your path. Remember, there is no rush to find this person. They will arrive when the time is right.

Life Path Number 3

If you have a Life Path 3, you are both creative and communicative. Public speaking, acting, journalism, or some type of art and craft is where your true talents lie. Your cheerful, optimistic, and highly positive nature endear you to those around you. Harmony, beauty, and the wonder of nature are areas that bring you great pleasure and peace.

You find it easier to solve other peoples" problems than your own, which means you often suffer quietly alone for fear of disclosing your own sadness. A warm, welcoming, and cooperative working environment is your area of choice. You are sometimes overly critical and demanding of yourself. Try to treat yourself more fairly (Weist, 2019).

Famous People

Other famous people with Life Path 3 include Snoop Dogg, Shania Twain, and Christina Aguilera (Laine, n.d.).

Your Relationships

Your current relationship is fraught with many ups and downs, and you are struggling to find the balance. Your view of love is based on an intangible fairytale-like existence, so it is often skewed towards non-existent perfection. If you believe your partner to be compatible, work to find balance by learning to give and take a little. If this is not possible, gather your courage and move on.

If you are single, put your unrealistic expectations aside and welcome potential new partners waiting in the wings. Learn to find the balance in your life without freaking out every time things don't go according to plan. Use your creative skills to create a more realistic environment whereby this new relationship can flourish.

Life Path Number 4

If you have a Life Path 4, you are a quintessential builder. Your trustworthy, solid, honest personality makes you one of the most sought after people around. You are idealistic yet practical, and you believe in the vision you create for your future. Your tenacity of spirit in working towards your goals is admirable, but remember not to overextend yourself. Learn to be a little more tactful and less stubborn (Weist, 2019).

Famous People

Sir Richard Branson, Brad Pitt, and Oprah Winfrey also share a Life Path 4 (Laine, n.d.).

Your Relationships

If you are already in a relationship, know that it is solid and well-grounded. However, don't allow yourself to become complacent because you are so comfortable and content. Find ways to bring a little excitement into the relationship by surprising your partner with a date night, a romantic weekend away, or an unexpected gift.

Life Path Number 5

If you have a lucky Life Path 5, you are progressive, eccentric, adventurous, yet relaxed. Your enormous compassion causes you to worry about those less fortunate than yourself. You will often put the welfare of others before your own. Monotony bores you to tears. You prefer lots of activities and enjoy motivating others. You will enjoy travel and a number of wonderful experiences during the year (Weist, 2019).

Famous People

Other well-known people with Life Path 5 include Steven Spielberg, Liv Tyler, and Jackson Browne (Laine, n.d.).

Your Relationships

If you are already in a relationship, you may find yourself stressing about feeling restricted. You will need to learn that every relationship involves some give and take, and a solid partnership is to be commended. Learning to compromise will bring greater depth and understanding into your relationship.

If you are single, this is not the year in which you will find the love of your life. Focus rather on traveling with a good buddy and use the year to enjoy yourself and fill your "experience bank" to the brim.

Wake me up When I'm Famous Signage. From Unsplash, Uploaded by Alice Donovan Rouse (n.d.),https://unsplash.com/s/photos/zodiac-signs

Life Path Number 6

People with a 6 Life Path are believed to be an "old soul," filled with wisdom and eons of learning. You are a natural nurturer and truly caring person on whom others rely. Your parenting skills make you a beacon of truth and kindness. You are stable and well-balanced, and prefer harmony to discord.

Life Path 6 indicates awareness and responsibility. You are happier at home where you're spending time with your family and community. Your loving, warm, empathetic, understanding, and compassionate nature draw people to you like moths to bright light. You get an enormous amount of pleasure and a sense of achievement from being of service to others.

Famous People

Other famous people with Life Path 6 include Albert Einstein, Goldie Hawn, and Victoria Beckham (29 et al, n.d.).

Your Relationships

Guard against losing your personality in your efforts to please others and focus on re--energizing yourself from time to time. Be aware of the fine line between caring and interfering, and make sure your offers to support other people do not become burdensome to you. You enjoy leading and guiding others to realizing their full potential; however, you sometimes neglect to allow yourself time to re-energize and recharge (Weist, 2019).

Life Path Number 7

If this is your Life Path, you are a perfectionist with deep spiritual insight. You have the power to heal or be a counselor or guide. As a leader, you use your intuition and experience to guide you. You are a loyal and devoted friend-,—always reliable and caring.

However, your emotional extremes create a see-saw effect in your life, which often disrupts your equilibrium and forces you into unpleasant situations, potentially causing you anguish and self-denigration (Weist, 2019).

Famous People

Princess Diana, Leonardo DiCaprio, and Julia Roberts are amazing Life Path 7s (Laine, n.d.).

Your Relationships

If you are currently in a relationship, you need to learn to discard your unrealistic expectations of your partner. You need to discover the difference between what you want, expect, and what you actually need. Your fairytale view of love does not provide a solid foundation on which to build a lasting relationship.

If you are single, do not rush into a relationship just for the sake of having a partner. Use this year to establish what it is you really want from a relationship. If you know you need something more than a distraction, you need to realize that commitment is required for a relationship to be successful. A partner who can validate you and offer you undivided attention may be a challenge to find. However, your patience will eventually be rewarded.

Life Path Number 8

If you have a Life Path 8, you are an industrious person who tends towards becoming a workaholic. You are career driven and although you may tend to have a quiet nature, you are an excellent leader. Your biggest downfall is that you tend to look to the material world for validation instead of looking inward. You tend to suppress your emotions, because you believe that making them known is a sign of weakness (Weist, 2019).

Famous People

Well-known people with Life Path 8 include Richard Gere, Whoopi Goldberg, and Stevie Nicks (Laine, n.d.).

Your Relationships

This is aa year to learn more about the true relationship you have with your partner. You need to really connect in order for you to discover hidden depths and exciting new aspects of your partner, as well as yourself.

If you are single, you are struggling to overcome the challenge of stepping out into the dating world. You are deeply afraid of being disappointed, so you tend to make excuses not to get involved in a relationship.

Remember, all successful relationships are built on trust and a deep connection. So, unless you are prepared to work at creating a great relationship, you will remain single.

Life Path Number 9

If your Life Path is 9, you are a highly emotional, humanitarian type of person. Your compassion for others runs deep, and your caring approach makes you ideal for becoming a teacher, writer, or philosopher. Although material success makes your life more comfortable, your main aim is to ensure that those in your care flourish and thrive. Guard against being selfish and allow yourself time off now and again to catch your breath and refocus (Weist, 2019).

Famous People

Other famous people with Life Path 9 include Mahatma Gandhi, Robin Williams, and Whitney Houston (Laine, n.d.).

Your Relationships

If you are in a relationship, you will need to learn to compromise. Bring back the romance into your life, and remember that there are two of you that need care and commitment. Don't let petty issues override your ego and dampen your love.

If you are single, then this is the year to dig deep into your own psyche to discover who you really are and what you expect from a relationship. Work at becoming that person you would like to be, so that when your soulmate arrives you are ready for a wonderful relationship.

Life Path 11/2

Your Personality

If your Life Path is 11, added to all the traits of 2, you are in an enormously beneficial position to positively influence others through your impressive skills. Your wisdom and knowledge is a source of light, power, and energy to those around you. 11s possess innate energy that others can tap into. You may not be aware of your power, though, and although you may be an introvert by nature, your inspiration and positive influence affect many people (Ganesha, n.d.).

Your Characteristics

Your soul is filled with wisdom, an abundance of knowledge, and truth. Your innovative approaches to problems ensure you will find viable solutions. Your presence fills a room and people try hard to emulate your skills and successes.

Your Career

You are a natural inventor and creative designer of new projects, bringing your wealth of skills, planning ability, and knowledge to every task you take on. Success is a way of life, although it comes at the cost of continued effort.

Hope Marquee Signage Surrounded by Trees. From Unsplash, Uploaded by Ron Smith (n.d.),https://unsplash.com/s/photos/zodiac-signs

Although your unique qualities make you stand apart from the rest of humanity, you are humble and self-conscious of your achievements. You never boast nor take credit for what is not yours. You set high standards for yourself and expect those around you to live up to your expectations. Although you are patient, you find foolish and wasteful behavior irritating.

Avoid standing in judgment of yourself and self-criticizing as these actions have negative implications for your mental health and well-being. Don't allow negative thoughts to infiltrate into your planning, as these disrupt your ultimate power. Careful use of your power will ensure you are able to do a lot of good for others instead of accumulating great wealth for yourself alone. Be aware of your healing power which can guide you into the field of medicine, acupuncture, or massage (Bender, 2019).

Your Love Life

You are sensitive to the needs of others and diplomatic in handling your relationships. You are a great lover, showing care, concern, and awareness of your partner's needs (Ganesha, n.d.).

Famous People

Famous people with Life Path 11 include Barack Obama, Tony Blair, and Michael Jordan (Laine, n.d.)

Life Path 22/4

If your Life Path is 22, you have all the traits of aa 4, plus the added advantage of having the skills to build structures, projects, and self-help schemes for the human race. Your structures may be huge enterprises that offer many jobs and training. Yours is the power to create amazing business opportunities that will have far-reaching effects on the community at large.

Your double 2 qualities strive for stability, a firm foundation, and harmony in your life. The qualities of 4 persist in the form of your desire for careful processing, rational thought, planning, and purposeful action. You avoid wastage in any form and strive to make use of everything you can to fulfill your goals.

Spiritually, you recognize and celebrate the fact that we are inextricably bound together through the cosmos, connected by a greater power-,—a supreme source of energy and love. This knowledge brings perspective, purpose, and meaning to your life.

Life Path 22 will lead to great wealth, and although you may reluctantly accept this fact because the 4 wants stability and doesn t like change, you will soon realize your full worth and power once you start out achieving your goals.

As per the power in your 22/4 Life Path, once you set up your enterprise, learn to train others and hand over the reins of responsibility to those who are capable of doing the job, your destiny will take you to a position of wealth and extreme success. However, try to lose your stubborn streak and learn to share the responsibility with others.

Because of your desire for harmony and control, you can become overbearing and demanding. Temper this desire by delegating tasks.

Learn to work with passion and dedication without working yourself into a state of exhaustion. Leave time for pleasure, and relaxation, and the opportunity to recharge your batteries (Bender, 2019).

Famous People

Other well-known people with Life Path 22 include Sir Richard Branson, Sir Paul McCartney, and Bryan Adams (Laine, n.d.).

Number 10

Ten is the significant number for peace and completion. Once a task is done, you will experience a sense of achievement and peace. All your hard work has paid off and now you have the opportunity to rest and recharge.

10s often enjoy spending time alone and allowing themselves the chance to relax before gathering their strength to focus on something new.

10s10s may experience new ideas, meet new people, or start a new project when they make themselves known. All the opportunities can re-energize your life and offer you the chance to open the door to increased positive energy (Bakula, 2016).

During the COVID-19 pandemic, many people were out of work. The number 10 became quite significant in my life, and so heeding the call to reflect and refocus was beneficial for me. Through a positive change in attitude, two friends and I started a small business making face masks. Very soon, we were able to employ a number of people who had been out of work. Eventually, we set up a number of small businesses in different districts that have proved very successful.

A Final Word

Once you have discovered your Life Path Number, remember that it will serve as an indication of the many facets to your character and in fact, your entire life.

Your Life Path Number impacts your health, wealth, career, love life, and relationships, as well as your destiny. So, as you can now see, your Life Path Number has enormous power.

Now let's look at the exciting role that numbers play in guarding your health.

Chapter 3: The Value of Numerology in Guarding Your Health

Though mathematics may not have been your favorite subject at school, the fact that it is a universal, constant, incorruptible, and faultless system of measurement in any form lifts it up to the realm of perfect mysticism. Math has the power to manifest all things spiritual in their purest form, thereby making them attainable and visible to you and all humankind.

Amazingly, all numbers are created from the ten basic numbers you learned in kindergarten. Each number has its own individual, divine magic and personal energy field. No two numbers are identical. Together, all numbers incorporate our entire universe and every manifestation therein. Those numbers that relate to you personally have the power to affect your love life, your health and well-being, your prosperity, sole purpose, and destiny.

Your Sacred Numbers

For the most part, humankind uses these sacred numbers to make sense of their lives, find solutions to problems, and assist them in making decisions for their future.

Perhaps you are one of many who believe that the date and time of your birth will dictate your success or failure in this lifetime. What if I were to tell you that your birth was no accident, and that you chose that exact time to arrive on earth in your physical form? Though this statement may sound somewhat bizarre, if it is indeed true, then you are the controller of your own fate. A deep understanding and appreciation for numerology can assist you in finding your true path through the chaos here on Earth.

Through numerology, we are able to assign meaning and value to the energy around us as well as that manifested around every area in our lives. The power of numbers should never be underestimated. As on a road map, your numbers are the lights illuminating the correct path you should travel in order to achieve your goals, stay healthy, and become the successful person you are meant to be.

63

Numerology is of great value to your health and well-being. Your Life Path number will give you a clear indication of how to protect your mind, body, and soul against illness. It can warn you against potential problems, foods to avoid, and how to remedy illness, allergies, and potentially harmful situations in your life (Using Numerology to Improve Your Health, 2008).

Your Numerology Health

You may not have been aware of the fact that your numbers play a significant role in your health and vitality.

Each of the nine sacred numbers controls a specific part of your body. Once you understand the importance these numbers have towards your well-being, you dramatically improve your chances of successfully controlling your health and fitness.

How Your Birth Date Affects Your General Health

Each person's birth date has a specific effect on their health and well-being. It is important to make the best use of your growing knowledge of numerology to ensure you improve your chances of maintaining good health.

Birth Date 1

Dates: 1, 10, 19, and 28

If you are a number 1 person born on one of the above dates in any given month, you are generally a happy, healthy, well-balanced person.

Health Issues

This Life Path indicates your potential to suffer from heart trouble, problems with your circulatory system, and erratic blood pressure. You may also be prone to blood- -related diseases. People with Life Path 1 generally suffer from excessive stress and may also experience problems with their eyesight.

Health Management

- One should keep to a strict healthy diet low in fats and red meat

- Rest as often as possible
- Take regular exercise
- Meditate
- Practice Yoga

Beneficial Foods

All forms of citrus are good for you. Also include kimchi, sauerkraut, and pickles that improve your stomach acid and help to lower your blood pressure.

Honey has particularly valuable healing properties, and you can enjoy dates, raisins, nutmeg, ginger, and barley. Herbs that work in your favor include thyme, lavender, and chamomile. Drink apple cider vinegar to help balance your stomach acid. A cup of chamomile tea before bed will help to calm your stressed system.

Papaya and garlic regulate your stomach acid production. Always relax as you eat, and chew your food well.

Lemon and Ginger. From Unsplash, Uploaded by Kim Daniels (n.d.),
https://unsplash.com/s/photos/ginger

Avoid

Gastric acid is essential for digestion. Some foods simply increase the body's acid level, however, causing you great discomfort. Thus, it will be wise to stay away from all foods that will raise your stomach bile levels. Avoid eating processed foods, vegetables, and fruits.

Other Matters to Consider

Get yourself into a good exercise routine. This will help alleviate stress and burn up excess energy. Exercise also provides you you an opportunity to unwind and to to remove yourself from stressful situations.

Birth Date 2

Dates: 2, 11, 20, and 29

Health Issues

If you have Life Path 2, you may suffer from excessive nervousness and stress. You may also experience insomnia, anemia, and struggle with breathing-related illnesses such as asthma and struggle with high blood pressure.

Health Management

- 2s need to eat slowly and chew their food well
- Avoid rich, spicy foods
- Find time to relax
- Practice mindful meditation

Beneficial Foods

A diet that is rich in flax, cabbage, cucumbers, and turnips will best suit you. Bananas and melons are also a great source of nutrients.

Leafy greens, liver, and seafood will help combat the challenges of anemia. Beans, nuts, and seeds are also good additions to your diet.

Avoid

Where possible, stay away from all types of emotional conflict and any situations that increase your stress levels.

Other Matters to Consider

Because you are a highly sensitive and intuitive person, you are quickly aware of negative vibes and discord.

Find time to practice mindful meditation in order to re-ground yourself and re-establish stability in your life.

You may also find kick-boxing a useful way to rid yourself of pent-up frustrations.

Birth Date 3

Dates: 3, 12, 21, and 30

The 3s born on any of the above days are more likely to suffer from the following health issues:

Health Issues

If your Life Path is 3 you need to take good care of your chest, lungs, and your upper respiratory tract. You may suffer from asthma and be more prone to pneumonia.

Because you tend to overwork, you are likely to experience increased anxiety which can result in the onset of type II diabetes. You may also struggle with skin problems such as eczema, psoriasis, or acne. You may have a tendency for arthritis.

Health Management

- Avoid stressful situations
- Drink chamomile tea before bedtime
- Consume at least 2 liters of freshwater daily
- Avoid processed foods
- Avoid alcohol
- Cut down on your salt intake

Beneficial Foods

Foods to add to your diet should include asparagus, beetroot, apple, all types of berries, grapes, and peaches.

Mint, saffron, and olives will have a beneficial effect on your overall health. Try also poached salmon, kale, and red bell pepper. Foods like onions, cabbage, and apples help your body to increase its anti-inflammatory reaction to allergic compounds.

Person Holding a Green Apple. From Unsplash, Uploaded by Jony Ariadi (n.d.), https://unsplash.com/s/photos/good-health

Avoid

Stay away from all foods that have the potential to increase allergies. Included are garlic, fenugreek, ginger, and most dairy products. Avoid wheat products and alcohol.

Other Matters to Consider

Because 3s are more likely to suffer from emotional issues, they should develop a good exercise routine to keep their stress levels under control. They will also benefit from massage, reflexology, and mindful meditation.

Birth Date 4

Dates: 4, 13, 22, 31

If you are a 4 born on one of the dates above, you may experience health challenges in the following areas:

Health Issues

If you have a Life Path 4, you may well suffer from depression and bouts of severe anxiety such as agoraphobia, which prevents you from going about your daily business due to fear of crowded places.

Take care of your upper respiratory tract to avoid unnecessary colds, flu, and infections. You may also struggle with recurring urinary tract infections.

Health Management

- Avoid excessive dairy products in your diet

- Drink plenty of water
- Increase your intake of Omega 3
- Include walnuts, apricots, and apples in your diet
- Practice mindful meditation
- Take up an interesting hobby
- Seek the support of a professional if you are unable to manage your depression alone

Beneficial Foods

Indulge your palate with all types of green vegetables and fruits that are rich in carotenoids, minerals, and vitamins that boost your immune system. Included here are spinach, collard greens, kale, a variety of lettuce, and chicory. Eggplant and kohlrabi are also beneficial to your health.

Avoid

A change in your diet can make all the difference to your health, so avoid sugar and, wherever possible, all sweet foods. Try to avoid all confrontational situations and those that lead to arguments or any type of aggression.

Other Matters to Consider

4s prefer a solid, grounded type of life, where nothing untoward is expected to disrupt their balance and calm. Avoid becoming stuck in a rut. Take up a hobby and make a concerted effort to spend more time outdoors where your soul can enjoy the uplifting and healing qualities of nature.

Birth Date 5

Dates: 5, 14, and 23

If your Life Path is 5, you may find you suffer from mental strain and nervousness. You are also prone to upper respiratory tract infections including flu, coughs, and colds. Take care to guard yourself against kidney infections. You may struggle with insomnia and the resulting irritability.

Health Issues

5s born on one of the above dates may have to face one or more of the following health challenges:

- Excessive anxiety and stress
- Stiff muscles in the neck and shoulders
- Migraine headaches

- Insomnia
- Skin breakouts
- High acidity levels resulting in arthritis

Health Management

- Take up yoga or a martial art
- Practice mindful meditation
- Drink a cup of chamomile tea before bedtime
- Avoid excessive sweet and oily foods
- Cut down on acidity-making foods by adding more fresh fruits and vegetables to your diet
- Lean meat and fish
- Avoid citrus and preservatives

Beneficial Foods

Starchy foods like rice, potatoes, and oatmeal are beneficial for your dietary needs. Include mushrooms, macadamia nuts, and onions in your diet, as well as parsley.

Sweet potatoes, red bell peppers, and carrots are a great source of beta-carotene, which is highly beneficial for 5s.

Blueberries, red grapes, and cauliflower will help keep a good healthy balance for your kidneys.

Avoid

Stay away from sodium, phosphorus, and potassium. Cut down your protein intake.

As much as possible, avoid stressful situations that increase your blood pressure and hype your anxiety.

Other Matters to Consider

Free yourself from excess stress and responsibilities that weigh you down. 5s enjoy the freedom to make regular changes in their lives. Structure a few good routines for yourself and ensure you maintain a good exercise regime. Watch your diet and drink plenty of water. In order to improve your health, stay away from smoking and alcohol.

Birth Date 6

Dates: 6, 15, and 24

If your Life Path is 6, guard yourself against respiratory infections. Sinus and allergies are troublesome for the 6s. As a highly sensitive person, you may be prone to excess anxiety and stress. Women may suffer from breast-related-related health issues.

Health Issues

If you are a 6 born on any one of the above dates, although you may generally be a healthy, active person, you may find that you struggle with one or more of the following health issues:

- Infections of the ears, nose, and throat
- Some women may experience breast problems, including cancer
- Men may suffer from prostate problems
- Some anxiety
- Fertility problems

Health Management

- Guard against catching a chill
- Maintain a healthy diet and fitness regime
- Ensure you have regular medical checkups
- Take regular vacations or find time during your normal busy schedule for brief power naps

Beneficial Foods

Foods rich in vitamin C such as citrus, spinach, and pomegranates help to guard against flu infections. Figs, walnuts, and mint have added value for your diet.

Protein in the form of lean meats and fish are good for repairing body tissues that have been damaged by infections.

Turmeric and ginger have anti-bacterial and anti-inflammatory effects and are beneficial foods for fighting infections.

Avoid

Because of your highly sensitive and anxious nature, you should avoid sweet, spicy, and oily foods. To the best of your ability avoid stressful situations and step away from aggressive, abusive people. Avoid prolonged, strenuous activity.

Other Matters to Consider

Because of their desire to be in charge, sixes can easily fall into the "control-freak" trap. Find time to relax and learn to delegate chores to others. Take care not to binge eat when you become overly stressed or tired.

Take long, restful walks whenever possible to clear your mind and give yourself some much-needed breathing space.

Birth Date 7

Dates: 7, 16, and 25

If your Life Path is 7, guard yourself against indigestion and general infections. You are prone to suffer from skin irritations and poor circulation. Gout and arthritis are also potential enemies to your health.

Health Issues

If you are a 7, you are often stronger mentally, though you tend to become easily irritated and can quickly fly off the handle at the smallest provocation. Your constant impatience makes you highly susceptible to the following:

- Skin breakouts
- Nervous breakdown
- An abnormal fear of infection from germs
- Indigestion and gout
- Bipolar disorder

Health Management

- Regular deep relaxation
- Try not to become obsessive over cleanliness

- Eat smaller meals more slowly
- Include roughage and fiber in your diet
- Seek professional support if you are unable to manage your mood swings

Beneficial Foods

Fruit plays an important role in your diet. Include watermelon, apples, and apricots in your meals. Flax, ginseng, and green tea will help to protect you against infections and clean your blood by removing the excess acidity that can cause gout and arthritis.

Yogurt, fennel, and papaya aid digestion and are easily absorbed into your system.

Avoid

Smoking and alcohol are particularly dangerous to the health of a 7. Avoid stressful situations and overworking by carefully planning activities and projects in advance. Don't be shy to ask for assistance or support when needed.

Other Matters to Consider

Allow time for self-indulgence, meditation, and relaxation. The need to regularly recharge your energy levels is imperative for your good health and well-being.

Birth Date 8

Dates: 8, 17, and 26

If your Life Path is 8, you are more likely to suffer from liver problems and issues with your intestines. You may also experience frequent headaches, pains, and swelling in your legs and feet. Many 8s also have dental issues and often lose their hearing early in life. There is also a high incidence of blood pressure and heart problems among 8s.

Health Issues

If you are an 8, you may find the ailments that plague you have continued for a long time before you seek positive intervention to resolve these:

- Rheumatoid arthritis
- Frequent headaches
- Digestive problems
- Disorders of the intestines and liver
- Some trouble with joints in the legs
- Problems with breathing
- Some heart problems later in life

Health Management

- Stop smoking
- Avoid pollutants
- Increase your intake of fatty fish
- Include broccoli, ginger, spinach, and garlic in your diet
- Drink plenty of water
- Avoid alcohol
- Exercise regularly
- Get plenty of good quality rest

Beneficial Foods

Eights should include sour apples, pears, and tomatoes in their daily diet. Nuts, sweet peppers, and sunflower seeds are of particular benefit for 8s.

Cruciferous vegetables, beetroot juice, blueberries, and cranberries contain good antioxidants for your liver.

Avoid

Canned and fast foods are particularly bad for 8s, as are all foods with excess oil. Avoid losing touch with friends and family. Your gregarious nature does not do well in isolation!

Other Matters to Consider

Remember you don't have to take control of everything! Learn to delegate and take some much needed time off for yourself. The world is unlikely to stop turning if you decide to take a break! If you are an 8, you may have a more serious view of life that can curtail your enjoyment and happiness. Try to relax a little more, and see the light--hearted side of life. Let go of past irritations and hurts.

Dates: 9, 18, and 27

If your Life Path is 9, you may have suffered from a variety of childhood diseases like chickenpox, mumps, or even measles. Nines are prone to respiratory infections, pneumonia, emphysema, chronic bronchitis, and lung cancer.

Health Issues

If you are a 9, you are more likely to suffer from one or more of the following illnesses:

- Kidney and bladder problems
- Fevers
- Childhood diseases like measles, chickenpox, scarlet fever, and German measles
- Sore throat
- Pneumonia

Health Management

- Drink plenty of water
- Keep your diet rich in phytonutrients, usually found in blueberries

- Include spinach, kale, cauliflower, and garlic in your diet
- Participate in regular exercise
- Have regular vaccination against flu and pneumonia

Beneficial Foods

Protection against free radicals is essential for 9s, who should include grapefruit, kiwi fruit, pumpkin, and cranberries in their diets. Garlic, onions, peppers, and leeks also have great nutritional value for 9s.

Avoid

Nines will benefit from avoiding all greasy foods, smoking, and when possible, exposure to second-hand smoke and air pollutants.

Other Matters to Consider

9s are prone to be very hard on themselves. They often take on more than they should, believing only they are capable of completing a specific task.

Regular meditation and yoga classes will help 9s to relax and find that sacred space in which they can regenerate their flagging spirits. Regular massage, physiotherapy, or even a visit to the chiropractor can help 9s reduce their stress levels and ease their tight muscles (Using Numerology to Improve Your Health, 2008).

A Final Word

Please note that you are unlikely to be afflicted with every potential illness listed under your number. However, you may be more susceptible to some of the diseases and health problems mentioned. By taking continual good care of yourself, you stand a better chance of not falling prey to a great number of illnesses.

Your health is a gift you should never take for granted. By being able to utilize the power of the information you have gained through understanding your numbers, you will have better control over your health and well-being.

Chapter 4: Priming Your Numbers for Success in Your Career and Business

Having covered a number of valuable topics thus far, it's now time to get into the serious business of your career. Maybe you're wondering if the career you have chosen is the best for you, or perhaps you're stuck in a dead-end job that you'd love to pass up for something better.

Your perfect career and business venture may lie in a single number that has the power to change your life for the better. Let's take a look at how you can make better choices in your working life that has the potential to bring you increased wealth and a sense of achievement.

The Power of Your Life Path Number

It's your Life Path Number that holds all the answers to your career and business options! Sounds crazy? Well, before you decide otherwise, take a look at the following information before making up your own mind.

Your Life Path Number is the most powerful number in your number chart. It not only gives you information about your personality, but as you discovered in the previous chapter, it indicates important information about your health and how to safeguard yourself against a variety of potential problems.

Now you can discover further powers of your Life Path Number, which can give you the heads-up on the most suitable career for you.

Life Path 1

As you are a natural leader and motivator who thrives on good planning and organization, you will more than likely enjoy a less restrictive working environment where your leadership skills can flourish. You tend to have innovative ideas that you would like to see reach fruition.

Your ego desires recognition for your achievements, and you enjoy the status that comes from being in control.

Potentially Suitable Careers

Being in charge of your own company might be your ultimate dream. Pursue a new business venture as an entrepreneur, graphic designer, freelance photographer, or innovative craftsperson. Don't discard potential jobs in the military, navy, or law enforcement. Because you have great communication skills, you may do well as a politician or public speaker.

Life Path 2

Because of your ability to maintain unbiased views and act fairly and justly in a situation, you may find arbitration your best choice of career.

You possess a great combination of social and negotiation skills. You may find you are the go-to person for disgruntled co-workers seeking to blow off steam or searching for sound advice about work issues. Your ability to listen carefully to other peoples' points

of view and incorporate their ideas into the bigger
scheme of things makes you a valuable team leader.

Potentially Suitable Careers

Your active, inquiring mind keeps you alert to all
possibilities. Combined with your genuine care for
others and your deep sensitivity, mediation is a
potentially great career option for you. Consider
becoming a lawyer, counselor, or doctor. Psychology,
teaching, or nursing may also appeal to your caring,
nurturing nature.

Life Path 3

Your natural artistic and creative flair, as well as your
high energy and skill with words, leaves others in awe
of you. You have good social interaction skills, and
enjoy working with other people.

Potentially Suitable Careers

As you thrive on variety and change, a career in
entertainment or travel may suit you well.

Your artistic flair may involve music, songs, and
instruments. You may, therefore, enjoy working in a

theatre or on television. However, journalism, advertising, or starting your own band may suit you better.

If you are more intellectually inclined, pharmaceutical or medical science may float your boat. Or perhaps you would prefer a career in psychology or psychotherapy?

Whichever choice you make, remember: you like change, and that your somewhat unconventional nature encourages curiosity and a desire to branch out from the norm.

Life Path 4

Your careful attention to detail and excellent organizational skills, together with your disciplined approach to life and work, will best suit a job that involves a routine schedule.

You are comfortable working in a well-structured environment where your skills will be appreciated. You are up for a challenge and pride yourself on your strong work ethic and your tenacity of spirit to see a job through to its conclusion.

Others look to you for assistance with their organizational short-comings, and you are always ready with practical advice and helpful ideas.

Potentially Suitable Careers

Engineering or financial planning may suit you extremely well. Your brilliant organizational skills will prove useful if you chose a career in project management. However, you will be just as successful as an accountant or a lab technician.

Life Path 5

You are an amazing communicator who loves to travel. You are easily bored, and enjoy consistent change and challenges that keep you on your toes.

Potentially Suitable Careers

Your ideal career will be one that involves traveling around the world, interacting positively with people from every walk of life. Marketing may suit your selling and persuasive skills very well. You would also make an excellent teacher, working with a diverse population of students who wish to learn a variety of things. You are drawn to factual information and your logical approach to life will stand you in good stead for a career in the sciences, computer technology, or medical research. Advertising, public relations, and freelancing are also potential careers that offer the diversity you crave.

For those of you for whom the nine-to-five routine is an absolute killer, there's always the option of a stunt person, firefighter, or high-risk construction worker.

Change, excitement, and constant challenge keeps you active and vibrantly alert, so an office job is an absolute NO-NO for you!

Beware of your potential to gamble. This doesn't offer a particularly consistent, lucrative, and suitable career for someone of your talents.

Firefighter in a Burning House. From Unsplash, Uploaded by Jay Heike
(n.d.), https://unsplash.com/s/photos/firefighter

Life Path 6

Your desire for unity and harmony, in addition to your
compassion and empathy for others, will ensure that a
career in healing, teaching, or caregiving will best suit
you.

You have a creative streak which, in association with
your desire to be of service to others, may define the
career that best suits you. You are a good, reliable, and

dependable team player who enjoys job security and consistency in your daily chores.

Potentially Suitable Careers

Your compassion lends you well to a career in nursing, counseling, occupational therapy, divorce or family law, or pediatrics. You may also enjoy being a social worker, or even working in the fashion industry as a designer or interior decorator.

Whatever job you turn your hand to will be a success, because by your very nature and commitment, you will always give your best efforts to your career.

Life Path 7

You are a natural dreamer who loves seeking the "why" and "how" in all things. Your flair with working alongside people in a caring, intelligent, and sympathetic manner means you are well-liked and trusted.

Potentially Suitable Careers

Your ability to analyze problems and discover the deeper meanings for many natural signs and the reasons why things work the way they do makes you an ideal candidate for a healer, astrologer, numerologist, or psychic.

As a detective, forensic pathologist, researcher, or as a general practitioner, you would also be successful. Make sure that the career you chose rewards you for your efforts!

Life Path 8

You are the philanthropist who desires power, ultimate authority, and control in your life. You are ambitious, enjoy public life, and play at the center stage. You firmly believe you are an expert in your particular field, and therefore you don't take kindly to having your motives questioned.

Potentially Suitable Careers

Corporate life suits you well. Your desire to be the CEO ensures you will work tirelessly and diligently to achieve your goal.

You are not a risk-taker, thus those that work with you will feel comfortable and secure with you as their leader. Your consistent, sound work ethic ensures tasks are completed perfectly and on time.

Your diligent and almost clinical approach to work will make you a great surgeon, pharmacist, or medical practitioner.

Other careers that are also suitable include banking, law, finance, accounting, or stock trading. You will make a great life coach! Take care not to sacrifice your personal happiness for your career.

Life Path 9

Though you are artistic and very creative, you may be aware of your healing skills and your keen desire to make a difference in society. Your humanitarian nature encourages you to worry about the health and well-being of those less fortunate than you.

Potentially Suitable Careers

Humanitarian professions, such as those connected with spirituality and religion, will best suit you. Your prominent leadership skills will make you a great teacher or a civil rights activist. Remain focused on your goal and don't allow your emotions to take control of your mind.

Painting, writing, and music offer potentially good careers for someone with your wide variety of talents. However, you may also like to consider following a career in research, social work, or becoming an immigration attorney.

Wooden Gavel. From Unsplash, Uploaded by Bill Oxford (n.d.), https://unsplash.com/s/photos/attorney

How to Make Great Investments with the Help of Numerology

Investing your hard-earned money in shares and capital investments may be a daunting task. The share market is a vast institution governed by numbers in the form of share prices that rise and fall, according to a number of external factors.

Well-known numerologist Sapna Khemani suggests that stock market investments can be done quite successfully using your date and year of birth to determine your destiny number (Khemani, 2019).

In other words, if you were born on the 8 January 2005, your destiny number will be calculated as follows: $8+1+2+5 = 16; 1+6 = 7$

If you believe numerology can give you the best advice for investing in the stock market and in shares, there are a number of important questions to ask yourself before you take the giant step.

First, study your destiny number and decide if, in fact, this number gives an indication of your potential success in investing.

Next, check your date of birth for your lucky sector. This will give you further information about your investment potential. Each destiny number indicates a different luck sector in which you can invest with a greater chance of success (Khemani, 2019).

Number 1 - Dates: 1, 10, 19, and 28 - Public Sector Investments

If one of the above dates indicates your date of birth, you should consider investing in the public sector of the stock market. This pertains to organizations that are government-owned, as well as services supplied by the government.

A good example of a publicly-owned company is one in which the government owns the majority shares. Many national and federal companies in the US fall within this orbit.

Number 2 - Dates: 2, 11, 20, and 29 - Pharmaceutical Sector

Your best option will be to consider a lucrative investment in any pharmaceutical company, such as Johnson & Johnson, Pfizer, Novartis, or Merck.

Number 3 - Dates: 3, 12, 21, and 30 - Banking Sector

If you are a 3 born on any one of the above dates, investments in the banking sector may be beneficial for your monetary gain. JP Morgan Chase Bank with assets in excess of 2 billion dollars, Bank of America with over 1.8 billion dollars assets, and Wells Fargo Bank with over 1.7 billion dollars asset value, are three of the best-known banks in this sector.

Number 4 - Dates: 4, 13, 22, 31 - IT Sector

4s born on one of the above dates should consider investing in the IT Sector. This is a vibrant and lucrative area in which to own shares. Apple, Microsoft, Amazon, and Samsung are excellent IT companies.

Number 5 - Dates: 5, 14, and 23 - Telecommunications Sector

All the 5s born on any one of the above dates should consider investing in telecommunications. Verizon, AT&T Mobility, and T-Mobile US are among the most successful companies in this sector.

Number 6 - Dates: 6, 15, and 24 - Automobile, Restaurant, and Hotel Sectors

All the 6s born on one of the above dates will do well to invest in the automobile and hotel sectors. The Wyndham Hotel chain heads the list at present. Choice Hotels International runs a close second place, with Marriott International in third place.

Number 7 - Dates: 7, 16, and 25 - Oil and Gas Sector

Number 7s should invest in the oil and gas sector. Exxon Mobil, Global Gas, Natural Gas, and Fossil Fuels are among the biggest US companies in this sector.

Number 8 - Dates: 8, 17, and 26 - Mechanical Engineering Sector

8s born on any one of the above days have a great opportunity to invest in one of a number of highly productive companies in this sector. Examples include the following: NASA, the Boeing Company, Lockheed Martin Corporation, and of course, the Ford Motor Company.

Number 9 - Dates: 9, 18, and 27 - Property and Real Estate Sector

9s born on any of the above dates will do well to consider investing in property and real estate. Here are just four of the top property and real estate companies in the US: Keller Williams Realty valued currently at just over 366 billion dollars, RE/MAX at 270 billion dollars, Coldwell Banker Real Estate valued at 240 billion dollars, and Sotheby's International Realty valued at just over 102 billion dollars.

Perhaps by taking your number into account before you decide to invest in the stock market, you will stand a better chance of making a viable investment that will bring you favorable earnings (Khemani, 2019).

Numerology and Gambling

Are you a gambler looking to use numerology in the hopes of improving your success at winning?

Because numerology is believed to be one of the most accurate concepts known to humankind, it can be applied in every walk of life——including in the casino.

Once again, your date of birth indicates the magic number that has the power to work in your favor on the gambling floor. Though numerology is considered a pseudoscience, its strategies can be used to determine the best time for you to place a bet as well as the most likely numbers to choose from. It is generally believed that when you utilize numerology strategies, you can substantially increase your chance of winning (Tom, 2020).

US Dollar Banknotes. From Unsplash, Uploaded by Sharon McCutcheon (n.d.), https://unsplash.com/s/photos/wealth

What is Your Numerology Strategy?

To begin with, you need to calculate your Life Path number, which we have already discussed, as well as your Fadic birth date number. So, for example, Paul's number is 6 because he was born on 2 May 1988. His Fadic number is therefore 2. Paul's most suitable days to place bets will be on Mondays and Fridays, because the Fadic week is calculated from Sunday to Saturday.

Gambler Fallacies

Avoid falling prey to the following myths, which encourage you to gamble on numbers that have either not been particularly common, or, those that have. This fallacy is a misguided and non-mathematical approach to take which is likely to ensure you lose rather than win (Tom, 2020).

Can you choose a lucky number with numerology? Why not? Numerology is considered a science-based approach to successful gambling. By calculating your Life Path number, you are able to discover important personal traits and details that should help you to improve your game results.

The Traits of Gamblers

Number 1

You are a gambler with an innovative, positive, and strong personality. As a leader, you are independent, fearless, and self-confident. You have a keen awareness of the intentions of others and can quickly discern a scam.

Number 2

If you are a 2 gambler, then you have a calm and balanced view of the game. You are seldom caught up in the emotional upheaval that affects other gamblers. You prefer a balanced, diplomatic, and cooperative approach to the game. Always ready to listen carefully and attentively to the betting procedure, you make your move only when you feel sufficiently confident to do so.

Number 3

You are a creative, communicative, and artistically-- inclined gambler who makes good use of your finely tuned intuition to play the gambling game to its fullest extent. You are often the most successful at the game, yet modest when you win.

Number 4

As a number 4 gambler, you are consistent, calm, and totally disciplined. You seldom go out on a limb or succumb to the rush and hysteria of the masses.

Number 4 gamblers are reliable. You will give sound advice to others and will never take advantage of anyone. You generally give the game a great deal of thought and play with control and planning.

Number 5

If you are a 5 gambler, you love change, versatility, and lots of excitement. You take chances that are often successful and are always optimistic.

Number 6

As a 6 gambler, you are a well-balanced, calm, and thoughtful player. You seldom go out on a limb and never risk everything. You make good use of your natural intuition to choose numbers that resonate with your inner balance and harmony.

Number 7

If you are a 7, you will gamble with wisdom. You will utilize your analytical skills to make the best decisions based on your spiritual awareness and beliefs.

Number 8

Your greatest ambition is to win at all costs. Your desire for abundance, good fortune, and prosperity holds you in its firm grip when you gamble. You are easily angered when you fail to win, but equally overjoyed when the cards or dice fall in your favor.

Number 9

If you are a 9, you will gamble with care and wisdom, seldom exceeding your limits and understanding when it is best to call the game quits and go home. Your compassion and care make you a thoughtful and careful gambler, who will never waste your time or money (Tom, 2020).

A Final Word

Although the advice and options suggested here are worth your consideration, your choice of career is ultimately up to you. Before jumping into the deep end, consider all your skills and options in order to make the best decision for your future.

Remember, careers can change. As you mature and life brings its constant change, your interests and needs may change too. So don't for one minute believe that once you have made a decision on a career that this is where you will remain until the day you retire.

Life is too short to waste time on inconsequentials, so grasp every opportunity that comes your way and give your chosen career your best effort.

Chapter 5: How Numbers Influence Relationships

Interpretation of Numbers

According to Pythagoras, the universe comprises different energy vibrations, each of which is assigned a specific number. Each number generates a specific energy field which can be interpreted in different ways. So, a number may be seen as positive, negative, or neutral. Depending on how the number is viewed and the situation in which it operates, it will in fact affect its energy field and ultimately its interpretation for your life.

It can, therefore, be conceded that energy can either be in a positive or negative field. This may have a direct impact on all aspects of your life, resulting in good and bad times, happy and sad moments, and successes as well as failures.

Interpretation is highly subjective and influenced by various ideas, thoughts, or situations. However, your own outlook on life is influenced and in many cases, controlled by your subconscious, which will directly impact the energy field of your numerology chart.

For the most part, you are likely to want to focus on the more positive aspects of numerology. However, it may be well worth your while to consider the influence the negative translation of your numbers may hold for you. In any event, to be forewarned is to enable yourself to be armed and ready to deal with the unpleasant side of life.

The Positive and Negative Aspects of Your Numbers

Take a quick look at the positive and negative aspects of the following important numbers and how these can be interpreted in your favor.

The numbers 1 through 9 embody the full range of human characteristics. Each number has its own specific qualities and together, all the numbers have value for every person's destiny and life path.

Number 1 - Unconventional, Idealistic, and Ambitious

The positive aspects of number 1 are associated with ego, forward motion, a pioneering spirit, good leadership qualities, self-reliance, rebirth, and renewal.

The negative aspects include being contemptuous, idle, and disinterested. Number 1 is also associated with loneliness.

Number 2-Gentle, Diplomatic, and Tactful

The positive aspects of number 2 include being sensitive, discrete, diplomatic, well balanced, and amiable.

The negative aspects indicate discontentment, fear, and being gullible. 2s also often feel invalidated and underappreciated.

Number 3- Playful, Innovative, and Creative

The number 3 is closely associated with divine creation. Its positive aspects include affability and talent, as well as being inspirational, highly motivated, and positive.

The negative aspects of number 3 lend themselves to carelessness, melancholy, and moodiness. Number 3s will benefit from mindful meditation to calm their overexcited imaginations.

Number 4 - Practical, Systematic, and Methodical

Number 4 is inclined to display earth energy that pivots around a solid foundation. Its positive aspects include being attentive, tenacious, logical, and practical.

The negative aspects of number 4 are that they can tend towards rigidity in their thinking, or becoming impetuous and obstinate. Number 4s enjoy taking risks.

Number 5 - Persuasive, Dynamic, and Versatile

Number 5s are defined by their freedom, amusing personalities, independence, and progressive lifestyle.

The negative aspects of number 5 include recklessness, deceptiveness, and impatience. Number 5s delight in constant action and change.

Number 6 - Harmonious, Protective, and Nurturing

Nurturing, empathetic, and supportive number 6s are also caring, agreeable, and homely.

The negative aspects of number 6 include depression, negativity, lack of trust, and understanding. Number 6s enjoy progress and harmony.

Number 7 - Spiritual, Contemplative, and Insightful

These highly intelligent people are the detectives in numerology. They are pensive, perceptive, neat, and display exemplary investigative qualities.

The negative aspects of number 7 include selfishness, hesitancy, and uncertainty. Number 7s are generally quick-witted, perfectionists, and natural skeptics.

Number 8 - Powerful, Generous, and Business-Orientated

Number 8s are determined, ambitious, efficient, systematic, and practical people. These people are goal-oriented and make successful leaders.

The negative characteristics of number 8 include irritability, stubbornness, and high levels of anxiety. They are often workaholics who set themselves impossibly high standards. Number 8s believe in giving back to the community and offering their support and expertise to assist others who are less fortunate.

Number 9 - Giving, Sharing, and Caring

Number 9 denotes an old soul who is patient, animated, and charitable. These people aim at reaching the highest level of consciousness and spiritual enlightenment.

The negative aspects of number 9 include thoughtlessness, and the possibility of becoming choleric and challenging.

Planes Doing aa Cloud Show. From Unsplash, Uploaded by Sean Alabaster (n.d.),https://unsplash.com/s/photos/nine

In most cases, only a single aspect of your numbers will suit you as an individual, and seldom will two people have identical aspects throughout their number chart.

Through understanding the meaning of your numbers, you will be able to meditate on each in turn in order to draw upon their power into your life (Dworjan, n.d.).

Master Numbers

The numbers 11 and 22 are considered Master Numbers, which are indicative of high intelligence and learning. People with these Master Numbers have a greater chance of success and achieving their goals. Neither of these numbers is reduced if they come up as a total in your calculation.

Number 11

11s have the power to heal themselves and others through their highly tuned psychic abilities. These people have reached a state of perfect balance and thus have been empowered to share their power with others in need.

Number 22

This number indicates the master builder in society. People with 22 have the ability to create enormous wealth and opportunities for others as well as themselves. These people leave behind lasting legacies. They are industrious, extremely creative and reliable, with a strong and focused attitude to life, wealth, health, and success.

How Do I Find My Relationship Number?

As you already know, numbers and letters vibrate with their own particular energy. Some attract, while others repel one another. How can you possibly know if that person you find so attractive is, in fact, the right partner for you? Once you have discovered your Relationship Number, you will be able to find a potential partner.

Your Relationship Number is derived from the name that you are known by. This name may not necessarily be the same as the one you were given at birth. You may feel more comfortable with your nickname, or you may choose to use your married name or your birth name. The important aspect for the calculation of your Relationship Number is to use the name you feel most comfortable with and are proud of.

Your Relationship Number reveals how you interact with others, what you believe you need from a relationship, the value you place on a relationship, the way in which you interact with people, and the type of relationships you will form.

Assign the correct corresponding number to each of the letters in your chosen name. Add all the numbers and then reduce the total to a single digit between 1 and 9. Refer to the chart below for the letter values.

Pythagorean Chart

1	2	3	4	5	6	7	8	9
A	B	C	D	E	F	G	H	I
J	K	L	M	N	O	P	Q	R
S	T	U	V	W	X	Y	Z	

Now that you have discovered your Relationship Number, you are all set to find out more about which numbers are compatible with yours.

For example: DENNIS JOHN SMITH = 4+5+5+5+1 = 20; 1+6+8+5 = 20; 1+4+9+2+8 = 24. Add the subtotals as follows: 20+20+24 = 64. Reduce your total to a single digit: 6+4 = 10 = 1.

Now, let's help you find your most suitable partner!

Relationship Number 1

You are a person blessed with charisma and charm. People are drawn to your warmth and beguiled by your personality. You enjoy a wide circle of friends, many of whom are no more than casual acquaintances. You are, however, somewhat reluctant in allowing people to get up close and personal with you.

For this reason, the number of true and valued friends you acknowledge can be counted on one hand.

Your love of being center stage in a relationship and your desire to be the pinnacle of your partner's affection make you seek someone who will always hold you as their idol and place you on a pedestal.

Your ideal partner needs to be someone who will always look up to you, treat you with love and respect, and prefer you to be the leader and sole provider in the relationship. You should get on well with people who are 1s, 2s, 5s, and 9s (Ward, 2020).

Relationship Number 2

If you are a 2, you are likely to seek a close, intense, and intimate relationship with your partner. Because of your high level of empathy, you will offer your partner a great deal of support, protection, and care. Your need to feel emotionally and physically close to your partner makes you want to court your partner in a traditional and romantic way.

However, when you feel insecure or unsure of a situation or person's response to your advances, you tend to withdraw and retreat into yourself.

Your ideal partner needs to be someone who appreciates your caring nature, but who will not take advantage of you. This person should also want to be romanticized and enjoy the special little details you believe are important in a relationship. Your ideal partner may come from any one of the following: 2s, 6s, 8s, or 4s (Ward, 2020).

Relationship Number 3

You are a happy extrovert who loves to socialize. An entertainer at heart, you prefer to keep your relationships interesting and exciting by ensuring you are never tied down to just one partner.

Your mercurial nature makes it challenging for you to settle down into a run-of-the-mill type of companionship. Constant change and new, provocative options appeal to you.

Your ideal partner will need to be someone who has the ability to keep you on your toes. This person should have endless energy, be full of pizzazz, and be extremely sexy. Your ideal partner is likely to be someone from 5s, 7s, 8s, or 1s (Ward, 2020).

Relationship Number 4

If you are a 4, you are dependable, practical, and disciplined. Your love for order, planning, and consistent behavior are likely to spill over into your relationships.

Your desire for peace and harmony in your life makes finding a suitable partner quite challenging. The ideal person for you will need to be gentle of spirit, but have an adventurous streak that will keep your love life spiced and exciting without going overboard. Your potential partner may come from either 4s, 6s, 8s, or 2s (Ward, 2020).

Relationship Number 5

If you are a 5, your cheerful, witty personality has people flocking into your orbit. You love change and constantly being exposed to new ideas, people, and situations. Working at a relationship 9 to 5 is not for you. There is no offer of security or stability from your side. However, with the right partner, you could reach great heights of happiness and have lots of fun together.

Your ideal partner should be able to keep you entertained, possess a great sense of humor, and give you the space you need to maneuver without trying to tie you down. You are likely to get on well with people who are 5s, 8s, 1s, and 9s (Ward, 2020).

Relationship Number 6

If you are a 6, you are definitely content when you make other people happy. Your fear of conflict and any form of disharmony keep you well-balanced and focused on keeping the peace.

Long-term stable relationships are your preference because these types of partnerships allow you to feel secure and protected. Your goal is to eventually find the ideal partner with whom you can settle down and build your own safe-haven in which to raise a family.

Your ideal partner should be a home-maker and someone who will thrive in a calm and happy environment. They should not be likely to upset the peace in the home by wanting to enforce their own will on you. Potential partners may be found among the 4s, 6s, 2s, and 8s (Ward, 2020).

Relationship Number 7

You are a dreamer with a vivid, romantic, and somewhat unrealistic imagination. Your idealized vision of love leads to many disappointments when relationships fall apart. People are drawn to your charming, caring, and sensitive nature. You find yourself falling in love easily, and often with the wrong type of person. These failed liaisons bring you great heartache because they seldom live up to your expectations.

The ideal partner for you needs to be a person who is as romantic as you, but also has their feet firmly planted in reality. They should be able to love and support you through challenging times while reassuring you of their commitment to you. You may find just the right person from either of the following: 5s, 7s, 3s, and 9s (Ward, 2020).

Relationship Number 8

Wow! If you are an 8, your desire for power and control will override your relationships causing these to disintegrate before they even get started.

By learning to relax and calm down, (you don't always have to be in charge to make things happen, you know!) you have a better chance at finding just the right person with whom to strike up a great relationship

Because you have serious trust issues, your partner may feel that you become overbearing in your desire to know their whereabouts at all times. In fairness to your relationship, you need to step back a little and realize that your expectations are, in fact, unrealistic.

Trust is earned, so you need to start working on this from day one. With care, understanding, and persistence your relationship has the potential to blossom into a long-lasting alliance based on a deep and committed love for each other. Potential partners may be found among the 6s, 3s, 2s, and 4s (Ward, 2020).

Relationship Number 9

Hi, number 9! You are an easy-going, relaxed, and highly loveable person. Your friendly nature results in you enjoying a wide circle of friends and a vast number of acquaintances. Your love of life and constant demand for change makes long-term relationships a challenge. You shy away from commitments that will tie you down, restrict your opportunities to discover new places and people, and that will, in fact, clip your wings.

The ideal partner for you is someone with a free-spirit, who shares your love of adventure and the sense of freedom to explore the world. This is someone who does not necessarily want to settle in one place and enjoy the home comforts of a traditional hearth and home. You may find just the right relationship blossoming with someone who is a 7, 2, 1, or 5 (Ward, 2020).

A Final Word

So now you will realize that there is a lot more to finding the best partner for you than just finding a pretty face or a great personality. Compatibility, trust, and the ability to communicate freely and honestly are highly prized qualities. The awareness of the interactions between the energy fields of specific numbers will guide you in your choice of the perfect partner.

Chapter 6: Numerology and Parenting

Parenting is a skill you learn only when you have children of your own. Though there are self-help books and heaps of well-meaning friends and family members who will offer advice, you have to learn to be a parent the hard way. This means you have to be willing to sacrifice your freedom and begin to share your time, care, love, and sensitivity with the newest arrival in your family.

Child-rearing doesn't happen overnight, and there is no single childcare manual that will exactly suit your needs and concerns for raising your child. By learning the right parenting skills for your child who will have their own personality, you will have the best tools for the job (Cusante, n.d.).

Knowing how best you can guide, nurture, and support your child is probably the most important information any parent can have. No matter how much you read and learn, you are seldom ready to cope with your child's individual quirks and behavioral patterns. Using numerology to ascertain your child's needs and characteristics can go a long way to favorably support your parenting skills (Bender, 2020 a).

The Value of Numerology in Child-Rearing

Despite the fact that you may know your child's sun sign, you may find numerology a delightful and entertaining method for figuring out your child's temperament. Through using this interesting pseudo-science, you will be able to determine your child's idiosyncrasies, personality traits, and inner motivations. The subtle impact of numbers in your child's life will also give you an indication of their strengths, weaknesses, and potential obstacles in their life, and how you can assist your child in overcoming these in a positive and constructive manner (Howell, 2019).

By understanding your child's personality, you can manage your parenting skills in a far more productive manner in order to raise a happy and well-balanced child.

Numerology can help parents discover their children's true personality and then adjust their child-rearing skills accordingly. Interestingly enough, it's the parent's numerology that is primarily important in understanding their reaction to raising their children.

Besides getting to know your child better by using numerology, you can develop a more positive and mutually successful relationship with your offspring (Bender, 2020).

Colored Name Cards. From Unsplash, Uploaded by Chuttersnap (n.d.), https://unsplash.com/s/photos/names

Parenting 1

If you were born on numbers 1, 10, 19, or 28 of the month, you will want to keep your child physically safe at all times——often without being aware of the importance of their emotional safety.

In order to be the best parent you can be, try to encourage your child to communicate their feelings and fears without reprisal from you. Allow your child to express their emotions and show your own love and affection for your offspring. There is no shame in demonstrating your love and care for your child.

You will interact well with children born under numbers 1, 4, and 9. Your relationship with these children will be harmonious and peaceful. However, you may experience challenges with children whose numbers are 6 or 8, as they are more emotional and moody.

Child 1

These children are creative, free-thinkers, and potential leaders. They may feel the need to try and overturn your rules because they believe they know better. Allow your child to forge ahead, but always be ready with words of praise as well as reassurance when needed. Also, offer clear guidance and advice on how your child can best deal with failure. These children thrive when they learn to believe in themselves.

Number 1 children are independent and enjoy demonstrating their prowess when it comes to trying things on their own. They will often tell you to leave them alone, or say "Let me do it!" Encourage this child to compete fairly and to accept failure with good grace. This child loves to be active and will not easily submit to staying still or being unable to get rid of their pent-up energy. A wide variety of sporting activities suit these children (Bender, 2020 a).

Parenting 2

For parents born on the 2, 11, 20, and 29 of any month, your parenting skills are usually highly developed because of the positive and nurturing relationship you enjoyed with your own mother.

Children born with birth numbers 2, 6, or 7 are likely to develop a more positive relationship with you. Those whose birth date adds to 8 or 9 may be more challenging and demanding.

Child 2

This child is a gentle, loving soul who will work hard to keep the peace, share toys and enjoy lots of hugs. You will need to ensure you develop this child's sense of diplomacy, harmony, and balance because this child was born to be a peacemaker. Be aware that this child is sensitive and caring. Give this child constant validation and recognition for all that they do.

Teach these children to become sufficiently strong to stand up for themselves without becoming someone else's 'doormat'. They need to learn that it is perfectly fine to say 'NO' without feeling obligated, or like they have somehow failed in their mission to offer support to others.

This child will adhere, without question, to the rules and boundaries you establish. However, they need to find suitable outlets for their emotions in order not to suffer blowouts or become depressed (Bender, 2020 a).

Parenting 3

Born on a 3, 12, 21, or 30 day makes you a 3 parent. You are sometimes torn between your work and family responsibilities, which can create unnecessary conflicts.

"Stick With me" Pink Neon Signage. From Unsplash, Uploaded by Cory Bouthillette (n.d.), https://unsplash.com/s/photos/zodiac-signs

Children with birth dates 3, 6, and 9 are most compatible with you. Those with 4 and 5 will challenge you for extra attention because they are more emotionally needy. You will require a great deal of patience to cope with these children.

Child 3

As a parent, you will need to help this child develop creativity and their ability to communicate their ideas, needs, and inner-most concerns. These children are afraid of criticism and are often filled with self-doubt. Listen carefully to what this child tells you and learn to read their body language in order to help them verbalize their fears and dreams.

Child 3 is bright, alert, and receptive to new ideas, though they seldom appear to be able to focus long enough to complete a task. Encourage this child to always finish what they start before they tackle their next project.

Parenting 4

If you were born on day 4, 13, 22, or 31 of any month, you are a typical 4 parent who will interact positively with children born under 1, 4, or 5. Being a moody person by nature, it will make you a challenging parent for your children to understand. Be sure to explain to them that your moodiness doesn't in any way mean you love them less.

Children born under 6 or 8 are more difficult to manage because they challenge you at every turn.

However, with consistent boundaries and loving support, these children will thrive.

Child 4

These children cannot be hurried as they become flustered and quickly lose faith in their own abilities. They thrive on lots of positive affirmations. Teach this child to take one step at a time and to learn the value of endurance.

Children who are 4s prefer stability, calm, and routine. They do not thrive in a chaotic situation as they quickly become frustrated and over-anxious. These children require added support in learning to express their emotions in a constructive manner. Their intense curiosity may lead them into danger, so keep a close eye on where these children are and what they are up to.

Parenting 5

If you were born on the 5, 14, or 23 you are a 5 parent who has a flexible, positive, and calm approach to parenting. Children born on the 5, 6, 9, 14, or 23 will fit in well with your personality, and your family will be filled with harmony, balance, and positive vibes.

Children born under 3 or 8 will be more challenging to raise, and you will need to be patient and understanding about their impulsive, busy, and loud personalities.

Child 5

These children need to learn to be self-disciplined and versatile in order to experience their true freedom. Guide your child to understand that they cannot experience everything all at once. Their adventurous, free spirit often leads these children into all sorts of challenging situations because they possess no internal awareness of the dangers in life.

Loving the spotlight as these children appear to do, often hides an underlying fear of failure. Help this child to realize their full potential by offering them consistent, emotional support, and clear, concise boundaries (Bender, 2020).

Parenting 6

Parents that were born on days 6, 15, or 23 have a naturally loving, caring, and nurturing approach to their children. For the most part, you will find that you interact well with all your children, although those born on 1 or 4 may confuse you at times with their quiet, withdrawn character.

Child 6

You will be wise to develop this child's sense of responsibility in every aspect of their life. These children often tend towards perfectionism and may display no real patience with themselves. It's an 'all or nothing' type of situation in which these children believe they are the only ones in charge. Their manipulative nature can create untold havoc if it is not controlled from the get-go.

Often precocious and outspoken, these children may appear cheeky and disrespectful without meaning to be so. They respond well to clear instructions which give them a good reason for having to complete specific tasks. They thrive in a solid, well-structured routine (Bender, 2020 a).

Parenting 7

Being a number 7 parent means you were born on the 7, 16, or 24 of any month. Your peaceful serene approach to life may be rudely disrupted by children not born under days 2, 6, or 7 whose boisterous natures create some confusion and disruption.

As a 7 parent, it is essential that you communicate your love and affection for your children in order that they learn they are loved and needed.

Child 7

These children are gifted with an awareness of a higher power. They are intuitive, deep thinkers whose analytical minds drive them to seek answers. You will need to be patient with this child as their mind is always elsewhere. They may be considered the "dreamers" in school who seldom appear to be on the same page as the rest of the class.

Children who are 7s don't trust easily and often end up feeling betrayed by friends and sometimes family. Encourage this child to be the best they can be, and never fail to validate their emotions as being real and of value.

138

Parenting 8

If you are a parent born on 8, 17, or 26 you may appear cold and hesitant in your parenting skills, as you may lack self-confidence. Ensure you communicate regularly with your children and develop a good and deep understanding of each other's idiosyncrasies.

"Please Stay on the Path" Signage. From Unsplash, Uploaded by Mark Duffel (n.d.), https://unsplash.com/s/photos/zodiac-signs

Children born under days days 3, 6, or 8 may prove easier to handle than others as they share many of your characteristics. You may find yourself frustrated by children born under 1, 4, or 5, as they will challenge you often.

Child 8

These children are intense and single-minded in their approach to gaining power, wealth, and control. They often have issues with authority and are often difficult to discipline. They are free spirits who are set on making their own successes in life despite their circumstances.

These children are often the achievers of the world whose goals are to reach great heights of wealth, success, and acclaim. Guide this child by setting firm boundaries. Train their leadership skills in positive ways so that they learn empathy and consideration for others.

Parenting 9

If you were born on 9, 18, or 27 you are a 9 parent, which equates to being brave, adventurous, and noble. These are the characteristics you will expect in your children. If your children are born under 1, 3, or 9 they will emulate your characteristics quite well. However, those born under 2 or 7 will not have the capacity to follow in your footsteps. You will need to nurture and support these children by showing them your love and acceptance despite their differences in character.

Child 9

This child has the wisdom and integrity many others lack. As someone with an 'old soul', this child is a humanitarian at heart who believes their duty is to serve the greater good of humankind. Their compassion is often used against them, so teach this child to be strong and to stand up for what they believe in without feeling remorse for being the amazing individual they are. Value this child's deep and lasting creativity and their inability to fit into the expectations of society (Bender, 2020 a).

A Final Word

Parenting is not an exact science. Many different personalities are involved in this area of your relationships, so it is realistic to expect some discrepancies in how you and your children relate to each other. Though you may find numerology helpful in giving you direction and advice on parenting, remember that every child is an individual in their own right and should be treated as such. Strategies that work for one child might turn out useless for another. The information supplied here is merely a guide.

Conclusion

Interesting Facts About Numerology

Now that you have discovered that you are surrounded by numbers in every aspect of your life, you may like to know more interesting facts about numerology.

Silhouette Photo of Four Palm Trees. From Unsplash, Uploaded by Adrian Trinkaus (n.d.), https://unsplash.com/s/photos/four

Positive or Negative Numbers

All numbers have both positive and negative features. Every number is affected by other numbers and how they relate to one another. Each of your special numbers will interrelate with each other and, in some instances, depending on your view of life and situation, may have a more negative effect on you.

In order to gain the best insight into your numerology forecast, you should be aware of the two opposing sides of your number. Life is not always filled with joy, happiness, and peace. Thus you need to expect some challenges along your life's path. Sound knowledge of numerology can, however, assist you in overcoming some of the negative aspects in your life and teach you to develop the positives in order to become happier, more successful, and to attain inner peace.

Numbers That Guide Your Life

In order to use numerology to its fullest potential, you should immerse yourself in its patterns and power. Many people advocate for the potential capability of numerology in bringing about major positive changes in their lives. When used correctly, numerology can help you achieve great success in mastering your time in order to create the peace and prosperity that you may only have dreamed possible.

Converting to Numerology

After reading this book you may find your appetite for numerology whetted to the extent that you decide to pursue further knowledge and a deeper understanding of the subject. Only once you have discovered the truth behind numerology and its power to change your life for the better will you choose to make it your own and grow your inner peace and prosperity.

Realization of Your Strengths and Weaknesses

Numerology possesses the power to disclose your weaknesses and strengths so that you may make timeous corrections and adjustments to yourself and your lifestyle. Your greater purpose in life will be disclosed, and you will find yourself less inclined to waste valuable time and resources on those things that you now know to be worthless.

Avoid Confusion

The combination of numbers that hold priorities for your destiny and happiness are varied and potentially confusing. It will stand you in good stead to seek the advice of a professional numerologist to help you interpret all the information your numbers will yield.

Numerology can be more valuable and insightful than astrology. However, combining both pseudosciences can create powerfully beneficial opportunities for you that you may never have expected.

Numerology Affects you From Birth

The exact date and time you entered the world plays a significant role in how your life eventually turns out. With the added knowledge and skills you have gained through studying and practicing numerology, you have a better chance of redirecting your life's path along the route to increased wealth, success, and happiness.

Numerology Brings Order

Discord and confusion are two of the factors that create disaster in your life, and are likely to lead to your failure to achieve your dreams and reach your goals. Through numerology, you can strive to reach harmony and balance-,—both essential qualities for positive growth and success.

Numerology and Quantum Physics

The new world order appears to suggest that all phenomena are interdependent. No single theory can exist in isolation. So, interestingly enough, numerology has an important role to play in the overall aspects of pseudoscience.

In order for the human spirit to progress in a positive and dynamic manner, it needs to be elevated to a higher level of consciousness. From there it will be able to realize its infinite relations to every aspect of the cosmos-, —without which it cannot survive.

It is therefore potentially valuable to note that although the human spirit and mind emanate from the physical body, they are interminably linked in order to forge a fuller awareness and understanding of the true value of life and its infinite possibilities.

The human mind has not yet reached a state of being fully able to comprehend the sheer magnitude of the universe and all its interrelated parts. Yet, we are aware of the orderliness of nature in order for it to continue its successful functioning (Decoz, n.d.).

Numerology and Unity

Numerology indicates an underlying unity in the universe as well as in your own life. Our date of birth and given name resonate with us in a way we may not fully comprehend. Simultaneously, every other birthdate and the given names of other people are inadvertently linked to our own, even though this may sound unbelievable. Naming someone or something is not an incidental act, but rather the result of a far deeper and intuitive action that holds a deeper meaning than we may be aware of. For example, naming a girl child 'Joy" resonates not only with inner happiness and an uplifting sense of joyousness but also refers to a deep sense of rejoicing, exultation, and victory.

Giving names to things and people according to our perceptions and understanding of the universe invokes all our senses, thus enabling us to experience the name in a variety of ways.

Therefore, the act of according a digital value to names, specifically birth names maintains that every person carries a name that is not only best suited to them, but that reflects their inner nature. Each name has its own perfect melody that synchronizes with the entire universe to bring balance and harmony to the one so named (Decoz, n.d.).

The Characteristics of Number

Every number epitomizes specific qualities that all people possess to a greater or lesser degree. Each number is considered an archetype and is individualistic in itself. No two numbers could ever be confused as each is so different from the others. For example, you would never confuse a 3 and a 7. Apart from looking different in their written form, they have dissimilar values and meanings. Each number can stand independently of the others, yet all numbers are interlinked and interdependent.

Each number has its own personality and once you understand this fact, you will be in a better position to forecast its response in any given situation.

Every human personality, quality, and characteristic can be deciphered by one or several of the cardinal numbers.

The nine archetypal human characteristics are represented by each of the nine cardinal numbers. All nine numbers, like our personal DNA sequences, are arranged in a different way in each and every person (Decoz, n.d.).

The number 9 is the symbol for returning to the starting point after the completion of a cycle. Interestingly enough when 9 is added to 5 the result is 14. Adding the 1 and the 4 brings you back to 5. Think of a seed. In nature, this cycle begins with birth, develops into maturity, and then falls back into itself in death, only to be reborn.

What Each Number Reflects

The number 1 is built like an upright pillar: Straight, tall, strong, and independent.

The gentle, humble, and almost subservient 2 is resilient, and can rise due to its spring-like quality after being crushed.

The number 3 is considered the most imaginative of the cardinal numerals. Creative, imaginative, and filled with inspiration, number 3 encompasses everything in this world and the world beyond.

The shape of the 4 designates a down-to-earth, firmly rooted number that designates determination and discipline.

The most dynamic number is 5, which pivots around a central point and hides nothing.

The most loving and generous number is 6, considered as the parent number.

Number 7 is accorded to the seeker of truth. It is a symbol of deep thought and wisdom.

The 8 designates balance between all things physical and spiritual. Its two well-balanced circles represent heaven and earth.

Number 9 completes the circle and like 6, is filled with love, but for the world more than the family (Decoz, n.d.).

Now that you have discovered the value of numbers for yourself, go in peace, find your inner strengths, and build on these to become the best person you were meant to be.

Bluesource And Friends

This book is brought to you by Bluesource And Friends, a happy book publishing company.

Our motto is **"Happiness Within Pages"**

We promise to deliver amazing value to readers with our books.

We also appreciate honest book reviews from our readers.

Connect with us on our Facebook page www.facebook.com/bluesourceandfriends and stay tuned to our latest book promotions and free giveaways.

References

29, et al (n.d.). Life Path Number 6. https://seventhlifepath.com/numerology/life-path-number-6/

Bakula, J. (2016, July 08). Numerology Influences Your Life. https://exemplore.com/fortune-divination/Numerology-Influences-Your-Life

Bender, F. (2020 a, May 19). Early Childhood Parenting With Numerology - Your Child's Life Path Number Explained. https://numerologist.com/numerology/early-childhood-parenting-with-numerology/

Bender, F. (2020 b, January 11). 2020 Numerology Forecast for 1 Personal Year: The AstroTwins. https://astrostyle.com/2020-numerology-1-personal-year/

Cusante, J. (n.d.). Numerology and Parenting. http://www.professionalnumerology.com/numerologyparenting.html

Decoz, H. (n.d.). AN INTRODUCTION TO NUMEROLOGY. https://www.worldnumerology.com/introduction-to-numerology.html

Dworjan, T. (n.d.). Studying Numerology for Beginners. https://horoscopes.lovetoknow.com/Study_Numerology

Faragher, A. (2020, April 10). How to Calculate Your Life Path Number Using Your Birth Date. https://www.allure.com/story/numerology-how-to-calculate-life-path-destiny-number

Fibonacci Number. (2020, June 26). https://en.wikipedia.org/wiki/Fibonacci_number

Ganesha. (n.d.). Life Path Number 11 Personality, Career, Business, and Love Life. https://www.ganeshaspeaks.com/numerology/life-path-numbers/life-path-number-11/

Holly, H. (2016, October 17). How Numbers Influence Us. https://www.absolutesoulsecrets.com/true-psychic-stories/how-numbers-influence-our-lives/

Hurst, K. (2020, June 10). Numerology: What is Numerology and How Does it Work? https://www.thelawofattraction.com/what-is-numerology/

Kali, A. (2019, October 10). The Essentials of Numerology for Beginners. https://www.nirvanicinsights.com/numerology-essentials/

Khemani, S. (2019, June 14). WHO SHOULD INVEST IN SHARE MARKET ACCORDING TO NUMEROLOGY? https://sapnakhemani.wordpress.com/2019/06/14/who-should-invest-in-share-market-according-to-numerology/

Laine, N. (n.d.). Numerology: Famous People: Life Path Number: Oprah: Bono: Bill Gates: Sir Richard Branson. https://www.tokenrock.com/numerology/famous_numbers/

The Public Figures Who Believe in the Mystical Significance of Numbers. (2011, September 17). https://www.independent.co.uk/news/people/profiles/the-public-figures-who-believe-in-the-mystical-significance-of-numbers-5334255.html

Tom, H. (2020, February 15). The relevance of Numerology in Gambling. https://hiddennumerology.com/the-relevance-of-numerology-in-gambling/

Tommer, S. (n.d.). Kim Kardashian. https://www.celebrities-galore.com/celebrities/kim-kardashian/life-path-number

Using Numerology to Improve Your Health. (2008, November 14). https://astronlogia.com/numerology/health/

Ward, K. (2020, March 17). Turns Out, Numerology Can Tell You Who to Date. https://www.cosmopolitan.com/sex-love/a31226112/numerology-relationship-number-compatibility/

Weist, B., (2019, December 09). Here's What Your Life Path Number Is, And How It Will Affect Your Relationships This Year. https://thoughtcatalog.com/brianna-wiest/2017/01/heres-what-your-life-path-number-is-and-how-it-will-affect-your-relationships-this-year/

CPSIA information can be obtained
at www.ICGtesting.com
Printed in the USA
LVHW091516221220
674903LV00031B/512